SORTING IT OUT
Evaluating Data

Barbara A. Somervill

The Rosen Publishing Group's
PowerKids Press™
New York

DISCARD

For the boys

Published in 2007 by The Rosen Publishing Group, Inc.
29 East 21st Street, New York, NY 10010

First Edition

Editor: Joanne Randolph
Book Design: Elana Davidian
Layout Design: Julio Gil

Photo Credits: Cover © Rick Gomez/Corbis; p. 4 © www.istockphoto.com/Galina Barskaya; p. 7 © LWA-JDC/Corbis; pp. 8, 12, 20 Cindy Reiman; p. 11 © John-Francis Bourke/zefa/Corbis; p. 15 montage by Julio A. Gil, Photos © www.istockphoto.com/Smiley Joanne, www.istockphoto.com/Thomas Polen, www.istockphoto.com/Perry Kroll, www.istockphoto.com/Peter Galbraith; p. 16 © www.istockphoto.com/Chris Crafter; p. 19 © Tom Stewart/Corbis.

Library of Congress Cataloging-in-Publication Data

Somervill, Barbara A.
    Sorting it out : evaluating data / Barbara A. Somervill.— 1st ed.
        p. cm. — (Think like a scientist)
    Includes bibliographical references and index.
    ISBN 1-4042-3487-X (lib. bdg.) — ISBN 1-4042-2196-4 (pbk.)
    1. Science—Methodology—Juvenile literature. 2. Research—Evaluation—Juvenile literature. 3. Technical writing—Juvenile literature. I. Title. II. Series.
    Q175.2.S658 2007
    507.2—dc22
                                                2005035727

Manufactured in the United States of America

Contents

Scientists like to find out why and how things happen in our world. If you ask why a tennis ball bounces strangely on an uneven surface, you are thinking like a scientist!

The Scientific Method

One of the most important things you will learn in science class is how to follow the **scientific method**. You begin by making a **hypothesis**, or guess. Next you must **research** your idea and plan an experiment to test it. You will need to record your results in a notebook.

When the experiment ends, you will have the data, or facts, you need to prove your hypothesis. Now you must sort and **evaluate** that **information**. You will use the most important data in your report.

Let's say your teacher has asked you to do a science **project**. While playing tennis, you hit a ball that landed on a crack and **bounced** strangely. You decide to figure out why the ball bounced the way it did.

Evaluating Research

The first time you evaluate data for your science experiment is when you begin your research. You must choose trustworthy, or reliable, books, magazine articles, or places on the Web for **sources**. Look for information from **experts**, **universities**, or scientific groups. These sources can be trusted and give facts that have been checked before being printed. Do not stop at just one source of information. If several experts give the same facts, you can rely on the information as being correct.

While doing research, you look up "tennis ball bounce science" on the Internet. You find Web pages from sports companies and universities. These places give information about the science of tennis balls and bouncing-ball experiments. This is the kind of information you need for your project.

Using a computer to search the Web is a good way to find information on a topic. You must evaluate the pages you look at to decide if the information is true and can help you with your experiment.

THE ORIGINAL
MARBLE COVER-100 Sheets

NAME

WIDE RULED
ROARING SPRING PA 16673

Using labeled flags on your project journal will help you find information quickly. The flags will work like an index in a book. An index lists topics covered in the book and where they can be found.

Looking Over Your Journal

You have written plenty of information in your project journal. You included your ideas, notes, measurements, and activities. You should have done your experiment several times. If you did this, you can compare the information from each try. Comparing results is one way to begin evaluating data.

To begin making decisions about your results, you first need to **organize** them. You should start organizing project information by reviewing your journal. Keep sticky notes and a pencil by your side. Write "research" on some of the sticky notes and place them on the pages that have your research. Do the same thing for measurements, supplies, experiment steps, and any other important information. Your sticky notes will help you find data when you need it.

Too Much Information

What if your project journal is 73 pages long? Some information is useful. Some is probably not. For your report you will need to decide which information is important to share. This is part of evaluating, or making decisions about, your work.

Writing a **summary** is a good exercise to help you narrow down your data. A summary is a short review of what your results tell you. For example, in your journal, you may have written four pages about the steps you took to run your experiment. Reduce that information to fit on half of one page. As you do this, you will be thinking about which facts are the most important. This is evaluating data!

As you look over the data in your notebook, think about combining information that is alike. This is another way to narrow down your data. One way to combine information is to put it in a chart or graph.

I. Introduction
　A. Summary of experiment and results
II. Hypothesis
III. Step-by-step plan for experiment
　A. Materials needed
　　1. Tennis balls
　　2. Surfaces
　　　a. concrete
　　　b. wood floor
　　　c. grass
　　　d. carpet
　　　e. wood deck
　　　f. tile
　　3. Ruler
　　4. Camera
　　5. Notebook
　B. Experiment
　C. Plan to track results
IV. Results
V. Graphics to support results
VI. Conclusion/summary
VII. Bibliography of research materials

After bouncing the tennis ball on different surfaces and recording the results, you will need to get ready to write your report. *Inset:* This outline shows how you might list information you want to include.

Making an Outline

When it comes time to write a report about your experiment, you will need to continue to make decisions about your data. What should you include or leave out? You have reviewed and organized the data in your journal and made charts. Now you should make an outline that shows what information your report will cover and the order in which you will present it.

An outline lists the main **topics** and support for each one. The topics you marked on sticky notes in your journal can be the main topics for your outline. You should include an **introduction**, the hypothesis, supplies, activities, measurements, and a conclusion. Support for the topic would include any details that back up your results, including any charts you made.

Explaining the Variables

A key piece of data for your report is a section, or part, on variables. A variable is an element in the experiment that may be changed or stay constant, or the same. Be sure to include this section in your outline. The variables in this experiment are the materials, including the tennis ball and the surfaces on which the ball is bounced.

Every experiment should have a constant variable. In this case the tennis ball is the constant variable. It is the same in all the trials you run. Dependent variables are the parts of the experiment that changed. In this experiment they are the surfaces on which you bounced the tennis ball. By studying the variable data, you will learn whether your hypothesis is true.

You tested how the tennis ball bounced on surfaces such as concrete, grass, wood, and carpet. A tool, such as a ruler or tape measure, might help you get a basic idea of how high the ball bounces.

1.

2.

3.

16 These pictures show a sequence. Step 1 shows the woman swinging at a tennis ball. Step 2 shows where her arms and legs are after hitting the ball. Step 3 shows how her body looks at the end of her swing.

What Order?

As you create your outline, you need to decide in what order you will present your data. You need to be sure that the order you choose makes sense. It should help a reader understand what happened in your experiment. A good way to present data from an experiment is in **sequential** order.

You should begin with how you got your idea and your hypothesis. Next tell the reader your plan of action or the steps you took to run your experiment. After you write the steps, give the measurements or results. Then write a summary and a conclusion that wraps up your experiment. Use words such as "first," "second," or "next" to help the reader follow your steps.

Backing Up Your Data

When you began your experiment, you did research. You evaluated sources of information and chose data to support your ideas. Now you must evaluate your research again. This time you need to decide which research to include in your report.

You may have found some research helpful as you narrowed your hypothesis. However, you do not want to include every piece of information you looked up for your report. You should choose to include only the research that will help the reader understand your experiment and its results. The reader does not need to know all the information you found on tennis balls. The reader just needs to know why the ball bounces differently on different surfaces.

It is important to note the sources you use in your report. You cannot use someone else's ideas or words as your own. When you note a source in your report, this is called an endnote or a footnote.

To create a visual aid for your report, have a friend or adult take a photo as you bounce the tennis ball on each surface. Place a copy in your report to help explain your experiment.

Visual Support

After you organize your information, decide what **visual** support you can use to fill out your report. You may use photos, graphs, tables, or diagrams to present data from your experiment. A visual aid can be a simple, clear way to present data that will help your experiment make sense to a reader.

Go back to your project journal. Evaluate which information can be shown in photos or diagrams. Decide which data can be presented in a graph, table, or other graphic organizer. Use your outline to help you organize where your visuals will appear. Note on your outline which visuals you will include and where. When you have both your written and visual information organized, you will produce a better report.

What Did You Learn?

Learning to evaluate data as you do an experiment is an important skill. This is a skill that you can use in many ways. You begin to evaluate data from the moment you start your research and you continue evaluating as you work on your experiment. When the experiment is over, you will evaluate your results. As you make evaluations, you are deciding what information is important and what is not. You continue this as you write your report.

In the ball-bouncing experiment, you thought tennis balls would bounce best on a smooth surface. You tested six surfaces, looked at the data, and came to a conclusion. Tennis balls bounced most evenly on smooth concrete. You proved your hypothesis true.

Glossary

bounced (BOWNSD) Sprang back up after hitting a surface.

evaluate (ih-VAL-yuh-wayt) To weigh the importance of something.

experts (EK-sperts) People who know a lot about a subject.

hypothesis (hy-PAH-theh-ses) Something that is suggested to be true for the purpose of an experiment or argument.

information (in-fer-MAY-shun) Knowledge or facts.

introduction (in-truh-DUK-shun) A beginning part that explains what is going to follow.

organize (OR-guh-nyz) To have things neat and in order.

project (PRAH-jekt) A special job that someone does.

research (REE-surch) To study something carefully to find out more about it.

scientific method (sy-un-TIH-fik MEH-thud) The system of running experiments in science.

sequential (sih-KWEN-shul) Having to do with the order in which actions take place.

sources (SORS-ez) The places from which things start.

summary (SUH-muh-ree) A short account of something.

topics (TAH-piks) The subjects of a piece of writing.

universities (yoo-neh-VER-seh-teez) Schools of higher education that are made up of one or more colleges.

visual (VIH-zhuh-wul) Having to do with something you can see, such as a photo, graph, or diagram.

Index

Web Sites

Due to the changing nature of Internet links, PowerKids Press has developed an online list of Web sites related to the subject of this book. This site is updated regularly. Please use this link to access the list:
www.powerkidslinks.com/usi/evaldata/